Georgi Marinov

Evolution of Crude Oil Price Term Structure

Georgi Marinov

Evolution of Crude Oil Price Term Structure

Recent Developments and Practical Implications

LAP LAMBERT Academic Publishing

Impressum/Imprint (nur für Deutschland/ only for Germany)
Bibliografische Information der Deutschen Nationalbibliothek: Die Deutsche Nationalbibliothek verzeichnet diese Publikation in der Deutschen Nationalbibliografie; detaillierte bibliografische Daten sind im Internet über http://dnb.d-nb.de abrufbar.

Alle in diesem Buch genannten Marken und Produktnamen unterliegen warenzeichen-, marken- oder patentrechtlichem Schutz bzw. sind Warenzeichen oder eingetragene Warenzeichen der jeweiligen Inhaber. Die Wiedergabe von Marken, Produktnamen, Gebrauchsnamen, Handelsnamen, Warenbezeichnungen u.s.w. in diesem Werk berechtigt auch ohne besondere Kennzeichnung nicht zu der Annahme, dass solche Namen im Sinne der Warenzeichen- und Markenschutzgesetzgebung als frei zu betrachten wären und daher von jedermann benutzt werden dürften.

Coverbild: www.ingimage.com

Verlag: LAP LAMBERT Academic Publishing GmbH & Co. KG
Dudweiler Landstr. 99, 66123 Saarbrücken, Deutschland
Telefon +49 681 3720-310, Telefax +49 681 3720-3109
Email: info@lap-publishing.com

Herstellung in Deutschland:
Schaltungsdienst Lange o.H.G., Berlin
Books on Demand GmbH, Norderstedt
Reha GmbH, Saarbrücken
Amazon Distribution GmbH, Leipzig
ISBN: 978-3-8443-2487-7

Imprint (only for USA, GB)
Bibliographic information published by the Deutsche Nationalbibliothek: The Deutsche Nationalbibliothek lists this publication in the Deutsche Nationalbibliografie; detailed bibliographic data are available in the Internet at http://dnb.d-nb.de.

Any brand names and product names mentioned in this book are subject to trademark, brand or patent protection and are trademarks or registered trademarks of their respective holders. The use of brand names, product names, common names, trade names, product descriptions etc. even without a particular marking in this works is in no way to be construed to mean that such names may be regarded as unrestricted in respect of trademark and brand protection legislation and could thus be used by anyone.

Cover image: www.ingimage.com

Publisher: LAP LAMBERT Academic Publishing GmbH & Co. KG
Dudweiler Landstr. 99, 66123 Saarbrücken, Germany
Phone +49 681 3720-310, Fax +49 681 3720-3109
Email: info@lap-publishing.com

Printed in the U.S.A.
Printed in the U.K. by (see last page)
ISBN: 978-3-8443-2487-7

Copyright © 2011 by the author and LAP LAMBERT Academic Publishing GmbH & Co. KG and licensors
All rights reserved. Saarbrücken 2011

To my family

ACKNOWLEDGEMENTS

A large part of this work was carried out in the first half of 2009, while I was doing an internship at Litasco SA, Genève. I would like to thank the whole team of the Company for their welcome and support, and especially Mr. Ludwig Hachfeld for his advice and assistance with the work during this period.

Table of contents

1. Introduction
2. Evolution of crude oil market
3. Characteristics of crude oil price term structure
4. Main factors, influencing the crude oil price term structure
   - 4.1. State of the world economy
   - 4.2. Crude oil demand
   - 4.3. Crude oil supply
   - 4.4. Crude oil reserves
   - 4.5. Futures trading
   - 4.6. Geopolitical events
   - 4.7. US Dollar exchange rate
5. Why understanding contango and backwardation is important - implications for structuring a profitable trading strategy
   - 5.1. Contango structure
   - 5.2. Backwardation structure
6. Historical development of crude oil price term structure
   - 6.1. Crude oil before the creation of the futures market
   - 6.2. Crude oil after the creation of the futures market
     - 6.2.1. Reasons for the creation of the oil futures markets
     - 6.2.2. Crude oil price term structure in the 80's
     - 6.2.3. Crude oil price term structure in the 90's
     - 6.2.4. Crude oil price term structure in the first decade of 21st century
     - 6.2.5. Contango and backwardation revisited and the implications for the various crude oil market participants
7. Analytical study of crude oil price term structure
   - 7.1. Methodology
   - 7.2. Reasoning
   - 7.3. Results
   - 7.4. Use of results for strategic decision-making
8. Relationship between crude oil, crude oil products and other energy commodities
   - 8.1. The example of ICE Gasoil
   - 8.2. The example of Natural Gas (NG)
9. The price term structure, utilized as a tool for forecasting purposes

10. Final remarks and conclusions

References

1. Introduction

Crude oil is the most important commodity in the last century. But in the same time its price is so volatile that it can be compared to a heart diagram. As crude oil is an extremely politicized commodity and being also among the most important drivers of the world economy, an analysis of its price term structure and its development throughout the years is very important by any criteria. Only in the last year, we observed an unprecedented, striking course of the crude oil price and its structure. Few people have imagined the big surge in the summer of 2008 as well as the immediate free-fall just 1-2 months after that. Therefore a deeper look at the price term structure of crude oil is crucial for analyzing and understanding oil industry and oil price at least to the extent possible. The work will include a brief overview of the crude oil price term structure before and after the creation of the futures market and will strongly focus on the price term structure from April 2008 to April 2009, since this time period represents reasonable boundaries for an effective analysis – it is long enough to give sufficient data and check different hypotheses and in the same time it is short enough to allow for a consistent and focused comment on the development of the crude oil price term structure, without too many distractions caused by the amount of the information.

The work is structured as to give a basic overview of the development of the crude oil market throughout the years, later explaining the main characteristics of the crude oil price term structure, which would be useful to follow the historical development of the crude oil price term structure, distinguishing between the characteristics before and after the creation of the futures market.

The work continues with commenting the main factors, influencing the price term structure and the way they drive it in one direction or another. Later, the basic states of the price term structure are tackled and challenged to get a clear idea of what they actually imply. The core of this work is represented by the analytical study, which deals with a large amount of data, collected from the ICE and NYMEX futures market exchanges and which gives the inputs for deriving important conclusions and information, which if used properly, could be the main tool for successful and profitable strategic decision-making.

In the last sections, the issue is how crude oil drives oil products and their respective relationship. Critical analysis of whether crude oil price term structure could be used to forecast market movements is done, followed by the concluding section, which summarizes the main points and findings of the paper.

I have chosen to study the term structure of the crude oil price in order to understand and explain some of the strategies employed by the major market participant in the oil industry for a few important reasons.

First of all, it gives me the opportunity to consider a significant time horizon, which tries to capture all foreseeable market conditions and the continuous changes that happen daily. The term structure incorporates any relevant information, be it current and expected oil demand, OPEC behaviour, volumes traded or anything else that influences the crude oil price and market and therefore represents a rather objective picture on which to base an analysis. Since it incorporates mostly futures data, it captures the expectations of the professionals in the industry today for what should happen into the future.

Moreover, the term structure is the instrument which shows the reader not only where the market is going but hints a potentially successful strategy that would exploit any state of the market, whether bullish or bearish. It recognizes the financial interest in oil trading and any speculative opportunities that a contango or a backwardation structures could present.

Last but not least, I build upon previous analyses of the term structure in the past, some of which I reviewed in the paper. This gives me a solid basis for understanding and analyzing the actual crude oil price behaviour of today.

2. Evolution of crude oil market

As early as the 4th century AD, China is credited with drilling the first oil well at a depth of about 240 metres. During the 15 centuries that followed, oil wells were dug in many places all over the world and the main usage of the extractions was powering street lamps, cooking and paving streets (with tar – a product of crude oil) in the big cities of that time. The first oil refinery was built in Jasło (former Austrian Empire, now in Poland) in 1854-56 and the main refined product was kerosene. However, the real boom of crude oil is related to the introduction of the internal combustion engine in the beginning of the 20th century. In fact, it has largely sustained the demand for oil until today.

Oil trading in this period used to be exclusively over-the-counter, based on bilateral negotiations usually directly between the oil company and the individual or company customer. The normal way of buying oil was to bring a ship to a loading station at a port, where you had to declare how much you want to load. All the trade was on the spot market.

Another important development of the 60's that has left its mark on the whole oil industry until today was the creation of OPEC (Organization of Petroleum Exporting

Countries). Its first creators and first members were Iran, Iraq, Kuwait, Saudi Arabia and Venezuela. It was founded with the intent to coordinate its member states' oil policies and as a reaction to a US law that imposed quotas on Venezuelan and Persian Gulf oil imports. Considered one of the most powerful cartels in the world, OPEC has often been blamed for any bigger increase in oil price. And while there are substantial proofs that it was the world price-setter during the 70's and 80's, after the establishment of the oil market exchanges and as a result of their growing importance soon after that, the influence of OPEC on the world oil price setting is not so evident.

However, with the growing importance of crude oil due to its progressively increasing demand (mainly because of the huge transformation and development of the industry, e.g. new refining methods leading to new refined products, most of the military forces of the world powers switching from using coal to using oil, growing importance of automobiles, development of the petrochemical industry, etc.) the need for regulation and creation of official trading market emerged. The catalyser of the fast reaction of the major economies in this direction was the oil crises in the 70's – the oil embargo by the OAPEC (Organization of Arabian Petroleum Exporting Countries – similar to OPEC, but including only the Arab oil producing countries of the Middle East)) proclaimed during the Yom Kippur war and the oil crisis of 1979, caused by the Iranian revolution followed by the Iran-Iraq war the next year. These events resulted in an extreme distortion of the oil market, due to sharp cuts of supply, which in turn led to a strong surge of the oil price with worldwide economic and political consequences. The following chart illustrates the extreme situation and provides one reason to why the biggest economies of the time decided to intervene by creating an official market exchange.

Chart 1: Historical Crude Oil Price[1]

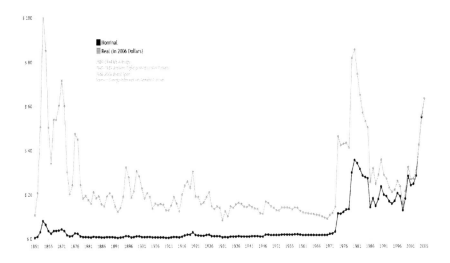

Source: Wikipedia

Chart 1 provides important information in order to understand the evolution of crude oil market. Since crude oil became an extremely important commodity in the second quarter of 20th century, stability of oil price in both nominal and real terms was observed. The striking increase of the price during the above mentioned crises (of more than 400% in real terms), provides a clear reason for intervention to try to escape similar situations in the future. With this intention, for the first time oil was being traded on the most important market exchanges at that time – IPE (specially created for trade in oil) and NYMEX.

With the launch of oil as a traded commodity on market exchanges and the introduction of oil futures, the oil industry has undergone a substantial change. Probably the most distinctive one was the initiation of trading "paper barrels". People began buying and selling oil futures, which irreversibly altered the functioning of the oil industry and added a great amount of new tools and policies available to all participants in the oil market, which have started using them with the idea of higher and/or safer profits. OPEC could no longer keep the price stable and its volatility increased tremendously, aided by the entrance of many non-oil companies (large banks, financial institutions, media) as key

---

[1] The data from 1945–1985 is the price of a barrel of light crude that was negotiated between parties "on the spot". The price from 1986–2006 is the price taken from the International Petroleum Exchange in London

players in the industry. This fact, coupled with some other factors, has led to the current situation where oil market has become among the most difficult things to predict. A bank, holding more than 2 million barrels (bbls) on an oil tanker is not uncommon. For example, in the beginning of 2009, Morgan Stanley and Citigroup put around 4 million bbls for off-shore storage with the aim to profit from the supercontango that prevailed in the market. Oil industry is heavily influenced by media, newspapers and reporting agencies, which are responsible for the setting of the world oil price and which quotes are used by all leading oil companies and institutions, including OPEC itself. The best example is Platts, a McGraw-Hill reporting agency, which employs various techniques to come up with daily updated prices for all relevant crude oil grades and other energy products, which have become the world benchmark, used by practically every big international company.

Naturally, the complexity of the oil market has grown exponentially. The current economic crisis has once again demonstrated the inability of any institution to control the stability of the market and the whole industry is seriously affected. After a differential of more than 100 $/bbl of the oil price between the summer of 2008 prices and just a few months after that, OPEC implemented measures to cut production in an attempt to recover the oil price to normal levels that would assure enough investments in the sector, urgently needed in order to sustain a stable level of supply.

However, it is yet unclear as to where the oil market is going. In the beginning of 2009, the leading international institutions in the oil industry – the International Energy Agency (IEA), part of the OECD and the Energy Information Administration (EIA), part of the US Department of Energy – forecasted in their monthly reports an average crude oil price in the range of 45-55$/bbl for the entire 2009. Under this scenario, many of the investment projects in the industry whether for exploration or for infrastructure were at risk because of the expected lower margins. Only two months after, we observed a steady price of around 70$/bbl and updated forecasts from the two agencies for an average price of around 60$/bbl. It remains to be seen in the near future as to what direction the market would take.

To summarize, the oil market's relevance has largely outpaced its alleged function to effectively match the demand of oil with the supply and provide an effective system for both oil producers and oil buyers to get a fair price. The entrance of various players, with the sole idea of profiting from speculation or trying to "outsmart" the market and never actually owning or delivering physical oil has distorted the fundamentals and has created a new, certainly more "hostile" market environment. Of course, one cannot forego the fact that the new players have had a positive influence by constantly creating liquidity and matching positions, which otherwise may not have been matched. But in times of crisis, when liquidity is most needed, they are the first to hide and pass the responsibility for fixing the situation back to the supply side. It is an important trade-off whether the oil market would have been better off without these players. It is interesting to note that in the summer of 2008, the US Congress has initiated investigation regarding the role of their largest investment banks in the record oil prices and in the oil industry market in general. The Democratic lawmakers criticized the lack of regulation for complex financial instruments and structured products as an explanation for the surge in oil prices. They were backed up by some large oil consumers (i.e. airlines), which complained that "speculators" in the over-the-counter market for derivatives (which is unregulated and therefore subject to a much higher risk), which were often investment banks, were pushing the prices of exchange-listed oil futures up. However, the ruling party defended the banks and dismissed the statements, qualifying them as invalid.

Whatever the truth is, the variety and number of oil futures market participants would hardly decline in the decades to come. The high volatility would always be a "magnet" for financial companies and individuals that pursue quick profits by betting on nothing more but subjective forecasts and expectations. But with the apparent consequences of the current economic crisis, the need for stricter rules, structured regulations and monitoring of the compliance of the oil futures and oil derivatives markets (both regulated and OTC) is obvious. And it would be no surprise if soon these markets and all oil trading activities enter a new stage of their evolution and development.

3. Characteristics of crude oil price term structure

We can distinguish between spot and futures crude oil markets. The former, known also as the physical market, is related to transactions which include the sale and purchase of crude oil in the near-term. A typical example is a contract signed in March for delivery in April and always entailing physical delivery and taking of delivery. The common participants in the spot markets are producers, refiners, traders, transporters that transact throughout the whole value chain - from the oil well to the final consumer. These markets benefit the participants by allowing them to more easily adjust to the existing demand and supply conditions.

The crude oil futures market in contrast deals with sale and purchase of oil in a future period. A futures contract is a legally binding agreement that carries the obligation to deliver a pre-defined quality and quantity of crude oil at a specific location at an agreed point in time in the future at a price, negotiated today. Buying and selling of futures contracts is done on organized exchanges such as NYMEX in New York and ICE in London. The majority of the futures contracts are closed out (offset) before the final settlement date (by selling a long position or buying a short position, depending on the type of the futures contract entered) meaning futures transactions rarely lead to physical delivery. The main participants in the futures market are mostly traders, representatives of international oil companies, commercial and financial institutions (e.g. investment banks, hedge funds). The main benefits vary depending on the participant in the futures market. Vertically integrated oil companies use futures market mainly for hedging the risk of fluctuating oil price and guarantee a minimum price for their product. Commercial and financial institutions usually engage in complex transactions, involving diverse financial instruments (quite often swaps) and try to predict the course of the market with the incentive of huge gains if they are correct. Often they engage in speculative transactions, which however are able to distort the market fundamentals and create an artificial imbalance.

Going deeper into the structure of the market and having distinguished between the two types of crude oil markets, it is time to see on what many strategic decisions in the oil industry are based. Crude oil price term structure is characterized by two different and mutually exclusive states – contango and backwardation. It is these two fundamentals that describe the oil futures market and provide crucial guidance to the major participants. Enormous profits and losses are generated every day because of the prevailing structure and various tools and techniques have been developed to benefit from both structures – the question is to foresee which one would dominate.

Contango is the situation where futures prices are higher than spot prices. The difference between the spot and futures prices is called basis. In the case of contango structure, the basis is negative. Usually the basis would be different in each delivery month and also negative. This reflects the fact that normally the spot price is lower than the futures (see section 6.1).

Backwardation is the opposite situation, where futures prices are lower than spot prices. In this case the basis is positive. This situation is often considered abnormal by the industry participants although it has been the prevailing structure at least in the last decade. (see section 6.2).

I continue with an analysis of the factors that cause the oil market to be in contango or backwardation.

4. Main factors, influencing the crude oil price term structure

As one can expect, the conditions and events that cause the market to be in contango or backwardation differ a lot and are also quite diverse. Since oil has become such an important part of our lives and is considered the most important commodity nowadays, it is natural that there are many factors, which influence both its spot and futures price movement. The activity on the oil market is so big that only important macroeconomic events are able to push the market up or down. The data is not exhaustive of all functioning factors but I try to explain the most relevant ones. I start with the most generic, which affects every industry and economic activity, including the oil market and industry itself.

4.1. State of the world economy

A normally growing economy (absent any other distortion factors) usually creates the most common oil price term structure – a normal contango structure, which reflects that oil futures are priced slightly higher than spot price, considering they include normal costs of carry and time value of money (for implications – see the next section).

A "booming" economy on the other hand (growing much faster than expected) creates a situation of backwardation, where the spot prices for immediate delivery are higher than the futures prices. The situation is considered as abnormal since it signals insufficiency of supply. The classical example is the significant additional unexpected increase of the oil needs and demand of the big developing countries, while reporting double-digit growth. This is one of the explanations of the rallying oil prices in the autumn of 2004, when the China's, India's and Middle East countries' economies demanded substantially more than expected (the expected world demand growth for 2004 was 1.3 million bbl/d while the actual number was 2.6 million bbl/d), which could not be matched by an appropriate increase of supply, at least in the short term.

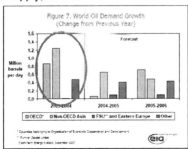

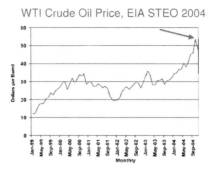

Finally, a slowing economy and/or an economy in recession are deemed to create a situation of a wide contango, the degree depending on the severity of the economic slowdown. This is explained by the (strong) fall in demand, decrease in the volumes traded on the futures exchanges and translates into a depressed world oil price at which fewer market participants are willing to sell. This is reflected in the oil price term structure by widening the contango and decreasing the basis, which sometimes even makes storage and futures sale of crude oil more profitable than the spot market.

In normal situation, the world economy usually does the "bulk of the job" of bringing the oil market in either direction. It works as to set boundaries in any single moment among which the price term structure can settle. And it is all the other factors that are responsible for the activities within these boundaries. A very good illustration is the state of the world economy in 2009 and the global recession that put enormous pressure on all countries and industries. The growth of the world economy ceased and was expected to be negative for this year and all this reflected on the oil industry by strongly contracting demand and a respective reaction of the supply. Since the beginning of the crisis, the oil price term structure has moved to a different path and has been "locked in" the attempts to limit the losses and recover the economy. Many professionals acknowledged that this would be a rather long period and various forecasts have been made for the behaviour of the oil price term structure in this time. In so far, lacking any major and unexpected distortion factors, the oil price term structure moves within the "invisible" boundaries set by the economic crisis.

There are some extreme cases of course, extreme factors different from the state of the world economy, which are able to significantly influence the price term structure. For example – the oil embargo and the wars in the 70's (geopolitical reasons), the extreme combination of tight oil market with growing demand and supply unable to match it in the autumn of 2004 (demand and supply reasons). But as striking as these are, they are the exception and not the rule, usually related to extreme deviations of what was expected, not subject to reliable forecasts. It is time to take a closer look at these factors.

4.2. Crude oil demand

The strong interdependence between the world economic conditions and the demand for oil would suggest that we are talking about the same effects they produce on the oil price term structure. However, this is not entirely true. Demand for oil falls when the economy is weak but up to a certain point. The modern society cannot function without oil and therefore a minimum "call" is always assured. By the same token, demand cannot rise indefinitely no matter a continuously bullish economy would suggest so. For example, there are simply capacity constraints, state and/or international regulations/interventions. All this to say, crude oil demand is able and do influence the price term structure in its own way. In most cases, it fine-tunes the price term structure and is responsible for its short-term volatility, also because in this time span, demand for oil is quite inelastic. A high price may shock and strongly affect users of oil but the commitments they made and their pattern of use of energy restricts them from immediate measures to shrink their demand. That is, they need some time to adjust. To illustrate, given the economic crisis during 2009 and its consequences, the demand for oil also left its mark on the price term structure in the recent period. The higher than expected decrease in oil demand in the beginning of 2009 contributed to the change in the price term structure, by pressuring the spot price and stimulating storage and buying futures contracts, assisting for a very wide contango. On the other hand, the strong and somehow unexpected level of oil demand from China, India and a few other developing (and "booming") countries in the second half of 2004 contributed to the price term structure, which was slightly in backwardation (Chart 2).

Chart 2

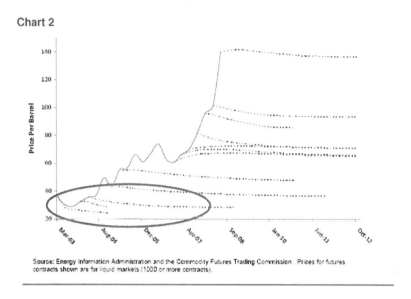

Source: Energy Information Administration and the Commodity Futures Trading Commission. Prices for futures contracts shown are for liquid markets (1000 or more contracts).

These examples illustrate how, although related to the state of the world economy, crude oil demand affects the price term structure and is a major factor to be considered and observed.

4.3. Crude oil supply

Crude oil supply influences the price term structure in a very different way and has been a sensitive variable even before futures exchange markets were created. While demand depends upon each country's specificities and usage patterns and in practice every country needs a certain quantity of oil every day in order for its normal functioning, supply is subject to each country's endowment with crude oil. In reality, 17 countries possess more than 92 % of the world oil reserves. The Energy Information Administration estimated a total amount of crude oil reserves of 1,332 billion barrels in 2008. Confronting this number with the top 17 countries by estimated oil reserves, we get 93.3%. The respective number for the OPEC oil reserves as a percentage of the total for the world is 70.64%. The data is represented in the following table:

Country	Oil Reserves, billion bbls
Saudi Arabia	267
Canada	179
Iran	138
Iraq	115
Kuwait	104
United Arab Emirates	98
Venezuela	87
Russia	60
Libya	41
Nigeria	36
Kazakhstan	30
USA	21
China	16
Qatar	15
Algeria	12
Brazil	12
Mexico	11.7

* Countries coloured in red were OPEC members in Q1 of 2009

Total (17 countries):   1242.7
Total OPEC:[2]   927.0

To summarize, the supply of oil is dominated by a small number of countries. Furthermore, more than half of these countries are united in the Organization of Petroleum Exporting Countries, which has been often proclaimed to be a dangerous cartel that influences the world oil price by adjusting its supply (sometimes unjustly and used solely to pass responsibility and deviate the attention from the real causes). Just from this data, the strong potential of the supply side to influence the crude oil price term structure is evident. However, soon after the creation of the futures exchange markets, the power of supply side has significantly deteriorated as well as some simple constraints have appeared. To justify this statement, I borrow some findings from Cynthia Lin (2009). She employed the basic Hotelling model in order to analyze historical data for world oil prices, oil price term

---

[2] Including the other 2 members – Ecuador (≈ 5 billion bbls) and Angola (≈ 9 billion bbls) – data from Energy Information Administration

structure and OPEC supply. She concluded that "the price term structure results suggest that OPEC producers exerted substantial market power over 1973-1990, but neither in the initial years of its formation nor in most recent years". While I will not go deeper into tracing the reasons for this since it is beyond the scope of this paper, it is a simple illustration of the relation between supply and crude oil price term structure and the potential of the supply side to influence it.

Similarly to the oil demand, the oil supply is, at least in the short term, also quite inelastic and therefore needs time to adjust to changing conditions like rising or falling demand, geopolitical events. In short – you do not just turn a tap to stop or start additional oil supply. That is, adding and cutting capacity is often time-consuming and expensive (operational costs, large amount of investments needed, contracts entered, laws of the country). That is to say that oil supply is a very effective factor and, again, is often responsible for sharp swings in oil price and oil price term structure. It is enough to recall the remarkable consequences of the oil embargo in the 70's, which led to an incredible distortion of the oil supply and the whole oil market respectively. The prices doubled even tripled in a very short period and if we had an established futures market to observe the oil price term structure, it is very probable we would have seen a quite wide backwardation. But let's return to realities and fundamentals.

In principal, when the market supply is adequate and covers the demand the situation is considered normal and, absent any major distortions, it creates a stable contango market structure. But quite often this is not the case. The fact that there are just a few major suppliers and the above-mentioned inelasticity of the supply, often lead to either undersupply or oversupply for certain periods of time.

For example, the consistent undersupply during the whole 2007 and the first quarter of 2008 strongly contributed to the gradually increasing oil price and reinforced the fears of lack of oil. This culminated in the end of the second quarter of 2008 when the market was reassured that there is enough oil by supply exceeding demand for the first time in a long period (the marked areas in **Table 1**).

Table 1

	2007				2008			
	1st	2nd	3rd	4th	1st	2nd	3rd	4th
Supply (million barrels per day) (a)								
OECD (b)	21.72	21.51	21.15	21.45	21.27	21.15	20.37	20.72
U.S. (50 States)	8.38	8.50	8.36	8.58	8.62	8.77	8.18	8.45
Canada	3.46	3.37	3.48	3.39	3.35	3.26	3.41	3.41
Mexico	3.59	3.61	3.46	3.35	3.39	3.20	3.14	3.11
North Sea (c)	4.81	4.50	4.29	4.58	4.47	4.33	4.02	4.15
Other OECD	1.49	1.54	1.55	1.56	1.53	1.58	1.62	1.59
Non-OECD	62.21	62.66	63.08	63.82	64.03	64.52	65.32	64.72
OPEC (d)	34.98	35.07	35.44	36.18	36.69	36.86	37.31	36.52
Crude Oil Portion	30.44	30.58	30.93	31.65	32.09	32.26	32.60	31.75
Other Liquids	4.55	4.49	4.51	4.53	4.59	4.60	4.71	4.77
Former Soviet Union (e)	12.61	12.60	12.55	12.66	12.60	12.60	12.43	12.79
China	3.92	3.96	3.97	3.96	3.93	3.99	3.96	3.92
Other Non-OECD	10.70	11.04	11.21	11.13	10.83	11.07	11.62	11.49
Total World Production	83.93	84.17	84.23	85.28	85.30	85.66	85.69	85.43
Non-OPEC Production	48.95	49.10	48.79	49.10	48.62	48.80	48.38	48.92
Consumption (million barrels per day) (f)								
OECD (b)	49.74	48.20	48.82	49.78	48.67	47.08	46.53	48.33
U.S. (50 States)	20.79	20.63	20.73	20.58	19.88	19.68	19.84	19.51
U.S. Territories	0.30	0.32	0.33	0.32	0.27	0.28	0.31	0.30
Canada	2.38	2.29	2.40	2.39	2.37	2.25	2.37	2.40
Europe	15.23	14.95	15.41	15.62	15.20	14.88	15.27	15.30
Japan	5.43	4.64	4.70	5.25	5.41	4.59	4.43	5.19
Other OECD	5.60	5.37	5.24	5.62	5.55	5.39	5.32	5.64
Non-OECD	36.11	36.68	36.72	37.16	37.40	38.18	38.24	38.56
Former Soviet Union	4.25	4.32	4.22	4.32	4.34	4.49	4.38	4.43
Europe	0.85	0.78	0.73	0.79	0.86	0.80	0.75	0.81
China	7.33	7.52	7.59	7.87	7.72	7.94	8.07	8.29
Other Asia	8.74	8.83	8.64	8.93	6.91	8.97	8.74	9.04
Other Non-OECD	14.94	15.22	15.54	15.26	15.57	15.98	16.31	16.00
Total World Consumption	85.84	84.88	85.54	86.94	86.07	85.27	84.77	86.90

Short Term Energy Outlook, EIA, December 2008

The whole scenario reflected also in the price term structure of crude oil and as could have been expected, the structure switched to backwardation as early as the second quarter of 2007. To put this formally, insufficient oil supply leads to some uncovered demand, a rising price of oil and creates fears of oil supply crisis, which rather soon are captured and assimilated by the market. The buyers want to insure against eventual continuation of the shortage of crude oil, which puts upward pressure on the oil price and makes it more desirable now than in the future – the exact conditions that transform oil price term structure to be in backwardation.

Exactly the opposite is true when we have a situation of regular oversupply of the market. The spare production that is not bought by the market accumulates either as an inventory or goes into storage and works as to depress the spot oil price and making futures much more attractive – therefore, creating perfect conditions for a contango structure (that could often be quite wide). Interestingly, after the period of undersupply in

2007 and the first half of 2008 and with the looming of the economic crisis and the evidence of continuous spare production in the market, the oil price started decreasing quickly, which also affected the price term structure and contributed to a new shift to contango after about a year of backwardation – another illustration of the influence of oil supply on the price term structure.

4.4. Crude oil inventories

Every state accumulates, among various commodities, also crude oil inventories. One of the reasons is to assure uninterrupted industrial activities in case of supply distortions for whatever reason (i.e. labour strikes in a supplying country, closure of delivery ports due to atmospheric conditions, interstate disputes). Another one is to take advantage of the low crude oil prices and build a higher level of cheaper inventories.
Whatever the reason, the level of world crude oil inventories is another important factor that changes the oil price and the price term structure respectively.
Theoretically, average level of inventories leads to a stable normal contango structure. Why? Normal inventories usually suggest that the oil market is working efficiently and is stable – the oil price is not low enough to stimulate building inventories and in the same time is not high enough to stimulate putting inventories on the market. Therefore, a normal price term structure is not surprising.

However, a qualification is needed. Inventories are not unique. They differ by their holder (i.e. commercial inventories, state inventories) and also by the way of holding (i.e. in-land storage, floating storage). This is important because sometimes it leads to opposite results to what the theory would suggest. For example, in 1991 total world oil stocks were at levels as high as 10 years before. The fundamentals suggest that this would put pressure on the oil price, which automatically translates into a contango structure. However, both the IPE and WTI NYMEX futures contracts were in steep backwardation. The reason was that in reality commercial stocks were very low but oil producing states were holding large amount of crude oil in floating storage near the major spot markets. Therefore, the supply tightened and the price term structure was instead in backwardation, in spite of the high levels of inventories. This is to signal that sometimes it is not enough to observe a single factor and its effects. It should be born in mind that a factor can produce different results depending on current market conditions. Furthermore, there could be an interaction between different factors with one prevailing over the other/s and - in the end it is the total effect of all factors that matter.

Logically, a low level of oil inventories usually signals a strong oil price, a strong demand/insufficient supply on the spot market and consequently a somehow backwarded market structure. The level of OECD stocks grew by more than 105 million bbls when we compare the average of the one year of backwardation (from the middle of 2007 to the middle of 2008) and just 2 quarters after that, when we observed a strong contango (Table 2). For the same period, on average the level of world oil inventories was consistently below the average historical levels (Chart 3). It is not coincidence that the term structure of crude oil followed closely and was in backwardation.

Table 2: Total OECD stocks (millions of barrels)

3Q2007	4169
4Q2007	4093
1Q2008	4099
2Q2008	4132
3Q2008	4178
4Q2008	4226

Source: International Energy Agency, Oil Market Report, 13 March 2009

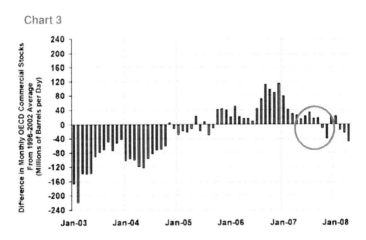

Source: Energy Information Administration: Short-Term Energy Outlook June 2008 and latest EIA data

4.5. Futures trading and activity on the oil derivatives markets

To demonstrate how relevant this factor is, I would make an analogy to the housing market in the US. In the past few years, it was widely believed that real estate represents a very profitable and safe investment. A large number of real estate agencies and individual investors were buying properties (without ever physically using them) with the intention to sell them in the future at a much higher price. This stimulated an artificially high ("false") demand for real estate (since the amount of investments increased tremendously), which naturally translated in artificially high prices. The worst affected were the people who really needed a home or a property but had to pay much above the fair price. We all know what the consequences of these inflated prices were.

This situation is very similar to what was happening in the oil futures and derivatives markets. Many companies, financial institutions and individual investors that buy "paper barrels[3]" cannot use them for anything since they neither own refineries, pipelines, nor have a place to store them. The thing they do however is speculate and hope the market would be favourable to their investment. I will hereby illustrate the serious role futures trading play in influencing oil price and its market structure. According to the information of the Oil Price Information Service (OPIS), on 6 June 2008 the contracts for oil that were traded only on NYMEX were more than 1 million, which equals more than 1 billion of crude oil barrels. And it is just paper that changed hands. For comparison, the oil actually produced on that day was approximately equal to 85.66 million barrels.

This is a relatively recent example of futures trading impacting the world oil price and its market structure. Bearing in mind the above numbers, a substantial role in the high oil price could potentially be attributed to the level of trade in futures. The proposed artificially high demand has been pushing the oil price for quite a while, reinforced by the attractive investment opportunities it offered, as high as it has been growing. And to support the fundamentals, the price term structure was in steady backwardation, which is evident from the following chart, representing the weekly WTI crude oil price term structure:

---

[3] A "paper barrel" is an oil cargo which is sold and traded on the open market, but not actually shipped

Chart 4

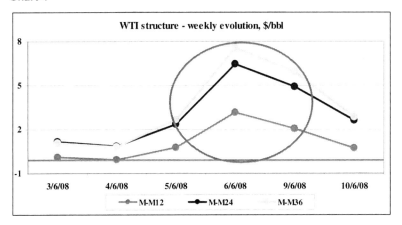

Source: ICE Database, Author's elaboration

Once again, the pitfalls of these huge trading activities were evident just a few months after the peak of the oil price.

Another effect of high levels of futures trading is that the market is misled as how much oil is needed in reality. The big activity on the market often gives wrong signals to the oil suppliers, which may misinterpret the real situation and bring more oil to the market than actually needed. This in terms could potentially reverse the trend, once the "bubble" of artificially inflated demand explodes. The market can become oversupplied, bringing quickly the oil price down and altering the price term structure.

To provide yet additional support to my arguments above, I use a previous work by Antoniou and Holmes (1995), who were testing the correlation between the volatility in the cash market and the introduction of futures trading. They found that the volatility increased after the introduction of futures, attributing this result to the role of futures trading in expanding the channels of information flowing into the market. And while the cash futures market is not identical to the oil futures market, it surely provides some important insights, which also apply to the oil market.

To conclude, oil futures trading and activities on the oil market exchanges are an extremely powerful factor, driving the modern oil market and price term structure. Oil futures could even be said to be the main factor in defining the oil spot price in the last few years. I make further comments on this issue later.

4.6. Geopolitical events

Geopolitical events are extremely important drivers of the oil price term structure since very often they have crucial effects on oil supply and demand and respectively the oil price. Any major conflict that involves an important oil producing country immediately creates pressure on the oil market and brings the price up. The two dramatic increases of the oil price that we observe in the 70's and the first years of the new millennium were both related to war conflicts involving major oil producing countries – the above-mentioned Yom Kippur war followed by the Iranian revolution and the Iran-Iraq war in the 70's and early 80's and the invasion of Iraq in 2003. Naturally, these translated into a strong impact on the oil price term structure, at least in the second example (table of term structure in 2003, showing wide backwardation – the circled area in **Chart 5**) since in the 70's and early 80's there was no market for futures trading. The dark blue curve represents the spot oil price movement for the selected years and the shorter curves represent the futures curves, which show the term structure.

Chart 5

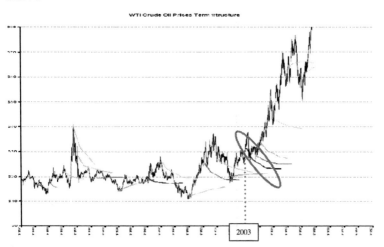

Source: John E. Parsons, MIT Center for Energy and Environmental Policy Research

But it is not only war or political conflicts that exert such influence. A devastating natural phenomenon or major physical accident (i.e. explosion on an oil field) could have similar effects, although usually concentrated in the short-term. For example, the Katrina hurricane that hit the US in the second half of 2005 had so serious consequences that they were passed to the world oil industry and oil market. Since US has been the biggest consumer of oil in the world and the tornado affected the core of its oil industry (the Gulf Coast where Katrina was most devastating supplied more than one tenth of the crude oil and almost half of all the gasoline consumed in the whole country), this had a worldwide reach. The following chart is illustrative:

Source: Short-Term Energy Outlook, EIA, September 2005

The crude oil price was rising even prior to the hurricane, but it nevertheless contributed to a mass panic mostly among gasoline consumers (used effectively by the retailers and wholesalers) who feared that the damaged refineries and the inelasticity of the supply (the imports increased, but it took at least 3-4 weeks until the carriers reached the US coast mostly coming from Europe) would cause shortages of gasoline. This scenario led to a sharp increase in both the crude oil and gasoline prices (mostly the gasoline price leading that of crude), which persisted in the next few months until the recovery of the affected region was effectively progressing and the imports were adjusted accordingly.

These examples are to show the potential of any major geopolitical event to significantly affect the whole oil market and its structure, whether in the short, medium or long-term. Usually it is the reactions of the countries' governments involved as to determine in which of these time periods the market distortion would fall.

It is simple to track the effects of geopolitical factors on the oil price term structure. As they are predominantly negative for the oil supply, they result in increasing oil price (sometimes very steep, depending on the severity of the crisis), which immediately reflect in the price term structure, which transforms into backwardation.

An example of a positive geopolitical event could be a warmer than expected winter, which weakens the demand for crude oil, lowering its price, which stimulates a stable contango structure.

Geopolitical events have been the major cause for most of the highest volatilities and biggest distortions of the oil market and its structure. They put an enormous pressure on the economy of all countries and create obstacles, especially in the short-term, to the normal day-to-day activities. In such situations, oil, which is a core input for many other industries of extreme social importance, could cost 2-3 times more than the previous day and this puts enormous pressure on all people. Therefore, it is of extreme importance to avoid when possible this type of events.

4.7. US Dollar exchange rate

The US Dollar is often preferred as the invoicing currency for international trading of commodities. And crude oil is not an exception. Therefore a fluctuation in US Dollar exchange rate is reflected accordingly in the volatility of crude oil price. When it depreciates, the nominal price in US Dollars of the oil commodity traded internationally should increase since more Dollars are needed to purchase the same quantity. Consequently, the weak currency should translate into a higher commodity price. In the analysis of the correlation undertaken by Professor Steve Hanke (2008), he found that if the US Dollar "had held its January 2001 value against the Euro, oil would have traded at about $76 a barrel in May 2008". However, this turned out to be $50 below the actual price in May 2008. An important result was that the decline of the value of the US Dollar throughout 2003-2008 accounted for the astounding 51% of the $97 per barrel increase in the oil price over the same period. Not surprisingly, from 2001 to the middle of 2008 oil rose approximately 160% in Euro versus 376% in Dollars. To summarize, a weak US Dollar tends to put pressure on the oil price and oil becomes more expensive. On the contrary, a strong and/or appreciating US Dollar makes crude oil cheaper and more affordable and brings its price down.

And while this is true, the extent to which the strength of the US dollar influences oil price and crude oil price term structure is somehow limited. This is also confirmed by Zhang, Fai, Tsai and Wei (2008). Using various econometric techniques among which co-integration, an ARCH type model and a VAR model, they conclude that there is a strong relationship between the US dollar exchange rate and the international crude oil price. They found out that the depreciation of the US dollar for the years from 2003 to 2008 was a key driver of the increase in the international oil price. However, their last conclusion is that compared to the powerful oil market, the influence of the US dollar exchange rate is proved to be partial. Nevertheless the strength of impact of US dollar exchange rate on the crude oil price, its importance for the oil market development and structure is doubtless. At least until it is no longer the world's most important financial reserve currency and the major one used in international trading. The G20 meeting in London during April 2009 tackled (unofficially) this issue of growing importance and may have actually given the start to the shifting away from the US Dollar. But it remains to be seen what the further development would be and until then, most companies would continue using the Dollar as the major invoicing currency for international trading.

5. Why understanding contango and backwardation is important - implications for structuring a profitable trading strategy

The crude oil price term structure represents a differential between a number of prices for futures contracts with predefined maturities. It is said to be in contango when futures prices rise with the maturities and in backwardation when they fall accordingly. The most basic illustration of why the term structure is important could be made with the following example. The profits made in the beginning of 2009 were possible thanks to regular monitoring and analysis of the term structure of crude oil. What it signalled to traders and investors was that the market was in such a wide contango ("supercontango") that in order to exploit it, they had to store crude oil and at the same time enter futures contracts for its sale after a few months which resulted in substantial margins and an almost risk-free profit. In practice, the regular and correct "reading" of the term structure could at any time give you important insights and information, which could be used in taking strategic decisions. Since usually contango and backwardation come in cycles, an investor or trader, while looking at the term structure is able to see whether it makes more sense to buy or to sell.

I continue with a closer look at the term structures of crude oil market, their occurrence and implications.

5.1. Contango structure

Contango is normal for non-perishable commodities, such as crude oil. The most trivial reason being these have a cost of carry. Such a cost includes storage fees and foregone interest of the money tied up in holding a commodity. Contango is usually present when supply exceeds demand. In this case, the surplus quantity is discounted to be sold more easily, reinforcing the contango structure, which in general is a characteristic of a bear market.

While I have already mentioned the main factors, which affect the crude oil market and their influence on its term structure, I did not comment on the implications of either of the two prevailing term structures. I use the next paragraph to tackle this issue.

Apart from rare situations of supercontango (extremely wide contango, creating time spread opportunities on the market – i.e. the futures prices of oil net of storage costs are much higher than today's), the contango term structure is usually normal – the futures prices comprise the spot price adjusted for the respective cost of carry, which produces a reasonable level of the basis without arbitrage opportunities.

The supercontango is relatively easy to spot and read – it implies a strategy of buying and storing (in-land or off-shore) crude oil at the spot price and entering futures contracts to sell the oil in (usually) the near future at a much higher price (exceeding the cost of carry) than it was originally bought. For example, in the table below there is a period when the differential between prompt and 1-year futures prices is more than $15, which having in mind the prevailing storage rates at that time, was large enough to make storing oil and entering into futures contracts attractive. The big differential allows for this almost risk-free profitable strategy. Almost risk-free since for example storage costs are dynamic and can increase enough to deteriorate the expected profits and inflict losses. Chart 6 illustrates this situation in practice.

Chart 6

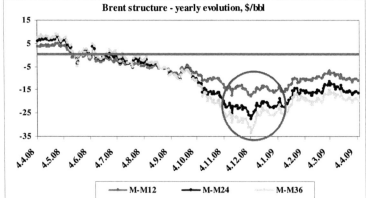

Source: ICE Database, Author's elaboration

The circled area in Chart 6 shows the period in which the contango was so wide that it allowed for the above-mentioned actions to be feasible. Considering the cost of storage and the opportunity cost of capital, it made economic sense to buy spot oil, sell futures contracts, hold it in storage for a few months and deliver when the maturity comes.

However, the market rarely allows for this situation and even if it happens, it does not last long. Therefore it is usually a normal contango when this type of term structure dominates. And it is much more difficult to read and build a successful strategy based on it.

What is behind the contango structure?

When trading oil, you inevitably have a time gap for allowing the oil to move from producers to consumers. Mentioned in the reasons for the creation of the oil futures exchange markets, time gap has been a big threat to the profit of refiners and to vertically integrated oil companies as well.

Figure 1

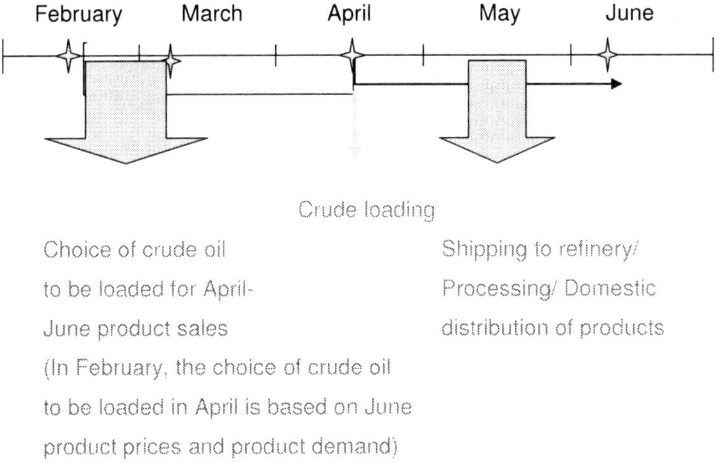

February   March   April   May   June

Crude loading

Choice of crude oil
to be loaded for April-
June product sales
(In February, the choice of crude oil
to be loaded in April is based on June
product prices and product demand)

Shipping to refinery/
Processing/ Domestic
distribution of products

Figure 1 above illustrates how spot crude oil market actually works. Oil producers decide whether and what products to produce in February, based on the futures prices prevailing for June. That is, the futures prices in June provide price guidance for what would be demanded in this period and the producers adjust their output accordingly. In April the producers enter into contracts for delivery, which is commenced usually more than 1 month ahead.

The implications of all this are tremendous. It means that spot prices of oil and oil products as well as the very term structure of the market are currently driven by the futures market and the trade on the major exchanges.

How contango and backwardation are created?

Especially in the last few years, the main oil producers and refiners have decided whether, what and how much to produce by looking in the future to see what the market would demand. The position that non-commercial traders[4] have taken plus the volumes of futures contracts traded in oil and different oil products have become the major indicator for the producers. And this is reflected into the price term structure. For example, the active trade in futures stimulates also the physical demand and is reflected also in the spot price and the price differential. When the futures price of an oil product rises and that of another falls, this signals to producers and refiners to shift producing more of the product with the higher future price in order to maximize revenue. And of course this affects the price term structure of both products widening/narrowing the backwardation/contango for the former and widening/narrowing the contango/backwardation for the latter. The same is true for crude oil and oil products. Falling futures prices of oil products show that there would be more crude oil on the market (since demand by refineries would fall) and therefore a fall in the crude oil price is very likely, combined with contango or narrowing backwardation.

5.2. Backwardation structure

Backwardation is the opposite of contango. It is very common in the perishable and/or soft commodity markets. Backwardation in the futures market usually occurs when there are supply inefficiencies in the corresponding spot market. In this case, the tight supply pushes the price for the physical commodity up, reinforcing the term-structure. It is common for bull markets in times of high economic growth and investment expansion. In the last section of the historical overview of oil price term structure I comment on the

---

[4] Companies and institutions which have neither the capacity nor the will to deal with physical crude oil and oil products and operate only on the futures exchange markets (i.e. banks, speculators, etc.)

reasons and conditions for the occurrence of backwardation as well as its implications for the main market participants.

6. Historical development of crude oil price term structure

We can begin talking about crude oil term structure right after the start of the trading of oil futures on commodity market exchanges. The first successful launch is traced back to 1983 when WTI Crude oil futures were established on NYMEX. Five years later, in 1988, IPE followed by introducing Brent Crude oil futures for trading. However, the term structure was not largely observed in the following decade and it was not until the entrance of various new non-oil participants in the oil market which brought an enormous increase in activity in oil futures trading that its regular monitoring became a must for anybody who wanted to be successful in this market. But let's start from the beginning.

6.1. Crude oil before the creation of the futures market

Initially (since the beginning of the 20th century), most of the official trade in crude oil was conducted under rigid long-term contracts, the basic case being a bilateral agreement to supply petroleum for a price and time period negotiated in advance. This resulted in a very low liquidity of global oil supply. In this period, the major oil companies were setting the crude oil price later substituted by the biggest oil exporting countries, which entered into agreements individually with consumer states. Although producers had to consider market forces such as supply/demand for oil and competition, selling prices were usually predetermined. Contracts for supply were concluded for long periods – usually 1-2 decades. All this led to an oil market, which was rather closed and rigid, unable to adjust to supply disruptions even in the medium to long term. Prior to the 70's oil embargo, however, few if any envisaged the structure of the oil market would create problems. Back then, it was in the interest of every oil producer to have concluded enough supply contracts so as to assure most or all of its exportable oil is sold virtually at the time of extraction.

Since the early 80's, the oil industry became increasingly dependent on the spot market, and consequently spot prices replaced the dominating selling prices. The relevance of the main oil exporting countries and oil companies also significantly diminished. And since crude oil futures were virtually non-existent, we cannot talk about term structure but only a single curve, representing the movements of the current spot price. Therefore it was only historical prices, which were quoted. And those were relatively stable, staying below $20 in inflation-adjusted US Dollars. It was not before the big oil crisis, caused by the war conflicts in the 70's in the Middle East and the oil embargo that followed, instituted by OAPEC that the spot price started fluctuating tremendously. These events, in reality, marked the end of the traditional oil market and signalled that soon significant changes would be made. And in fact, that is exactly what happened.

6.2. Crude oil after the creation of the futures market

The creation of the oil futures markets marked the beginning of a new era for oil and oil trading. The fundamentals changed and with them, also the influence of the old market participants. New players also entered, contributing to the increasing size and complexity of the oil markets. But let's take a look at the reasons for their evolution.

6.2.1. Reasons for the creation of the oil futures markets

Undoubtedly, the main catalyst of the appearance of oil futures and their introduction for trade on market exchanges was the oil crisis of the 70's. At that time, the US relied on imports for more than 1/3 of its oil consumption and the "targeted" embargo against it virtually blocked the supply of all of this quantity (most of the imports were from Middle East countries). Because of the above-mentioned characteristics of the oil market in this period, it was difficult for the US to find alternative sources for its imports, but in the longer term it became evident the instability and insecurity of the oil market, which had to be resolved. The severity of the situation at that time is illustrated by recently declassified British government documents, which revealed that both Britain and the US were planning to seize Middle East oilfields as desperate measures to counter the intolerable physical and psychological effects of the embargo.

Therefore, the creation of market exchanges for crude oil was largely a safety measure, guaranteeing continuity of supply even when there are market disruptions like the Arab oil embargo. It had to stimulate a shift from rigidity to dynamism, from a very low to a very high liquidity. It really did. And up until today, most of the crude oil is traded on these market exchanges.

It was also an insurance measure against a similar (post-embargo) rallying price in the future and represented the hedging solution most of the market participants desperately needed. Prior to the launch of oil futures on the market exchanges many oil companies fell in the situation of buying oil at the spot price, shipping it to let's say the US in about 45 days just to find out that in the meantime the price moved such as to destroy their revenues. It was bearable in the past when the oil price was stable and pretty linear but increasing volatility and uncertainty created the need for global hedging solution. And this solution was found in the institution of the futures market exchanges.

6.2.2. Crude oil price term structure in the 80's

Trading oil and oil futures on official market exchanges (in New York and London) was an immediate success in the '80s. For the first time there was room and opportunities for speculators to profit from buying and selling of crude oil. Exporters and importers had a global market, which they could easily use for many more and diverse tasks. The functioning of spot markets was also facilitated by the factoring of a wider range of market forces in determining a more objective and accurate spot price. New entrants and small companies had, for the first time, an opportunity to participate and do business on the market.

The increased activity, number and variety of participants and the introduction of new trading and financial tools naturally made the oil price much more volatile. The concept of basis and term structure were presented for the first time in relation to the crude oil market and with their growing recognition and importance they started being monitored and analyzed closely. It is time to take a closer look at the evolution of crude oil term structure. The data I found for this decade is very limited and starts from the middle of 1989. This is normal when we take into consideration the time when oil futures were launched (middle of 80's) and the understanding and entrance into usage of the concept of term structure. In the following chart, I represent some of the first recorded data of evolution of crude oil spot and futures prices, which later formed the basis for the crude oil term structure as we know it today.

Chart 7: Evolution of the spot and long-term prices of oil between 1 June 1989 and 28February,1991

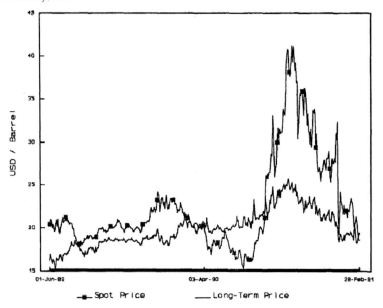

Source: Jacques Gabillon, The Term Structure of Oil Futures Prices, Oxford Institute for Energy Studies, WPM 17 1991

Chart 7 clearly shows that the spot price is higher than the futures throughout almost the whole period and therefore this translates into a term structure that is in backwardation.

The next chart (Chart 8) represents the volatility of both prices during the same period.

Chart 8: Spot price and long-term price volatilities from June 1989 to February 1991.

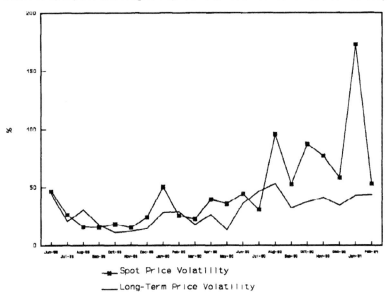

Source: Jacques Gabillon, The Term Structure of Oil Futures Prices, Oxford Institute for Energy Studies, WPM 17 1991

The new-born market for crude oil futures shows to have been much less volatile than the spot one, signalling that further significant growth in trading still remained to be seen.

To summarize, the 80's put the beginning of the new "oil order" with the institution of the crude oil market exchanges. It was also in this period that the concept of basis and crude oil term structure were gaining ground. After this, the road for the crude oil market was open for further evolution and developments, increase in complexity and enormous growth of its importance for the global economy. It marked another event that has to be accounted for – the powerful role of the US Dollar as the major international currency was consolidated with the creation of the liberalized crude oil market exchange, entirely denominated in US Dollars.

6.2.3. Crude oil price term structure in the 90's

The new decade witnessed surging demand for crude oil and a big increase of activity on the oil market exchanges. This also translated in the term structure during this period, with a steady backwardation, reflecting the pressure on the price that was put by the rising demand. This is represented with the first ellipse in the following chart:

Chart 9

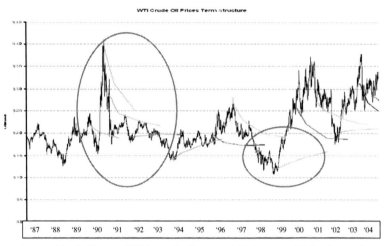

Source: John E. Parsons, MIT Center for Energy and Environmental Policy Research

Following the economic crisis of the previously booming Asian developing countries in the end of 1997 coupled with the increase of the OPEC production quota by 2.5 million barrels per day (despite the slowing of some of the important oil consuming economies) led to a sharply decreasing oil price. The marked areas in the below chart illustrate this scenario.

Chart 10

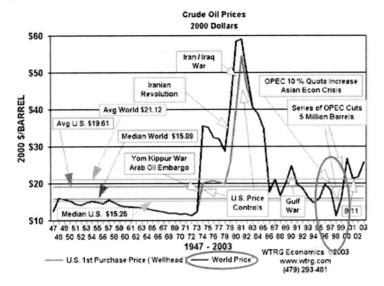

According to theory, what should have happened during this period of falling demand and oil price while supply was rising is a shift of the term structure to contango. And this is exactly what happened with a more or less constant contango from 1997 until 1999 (the second ellipse in Chart 9). OPEC reacted accordingly by correcting its quotas and supply plans by a series of production cuts, which in the end were successful in "reviving" the upward course of the oil price as soon as the end of 1999. And consequently, we observe a new shift back to backwardation.

In conclusion, the 90's presented some challenges for the oil market and its normal functioning, as well as tested the hypothesis that changes in oil price are soon reflected in the term structure of crude oil. The data analyzed and the results obtained clearly support this hypothesis and allow us to continue tracking the evolution of oil price term structure also for the decade we are living in.

6.2.4. Crude oil price term structure in the first decade of 21st century

The first years of the current decade were marked by the 9/11 attack, which psychological effects are still recognized even today, and the US-led invasion of Iraq. Both of the events were expected to have profound and long-lasting consequences on the oil industry and oil price. The former one affected only marginally the oil market by undermining the demand and price of oil, immediately counteracted by OPEC production cuts, but its "invisible touch" was in changing people's mentality all over the world and laying potential "traps" ahead for the valuable commodity (i.e. the increased sensitivity of people to information related to war conflicts and other distortion factors of the oil market significantly increased the volatility of the oil price and the amount of information included in it). The latter, on the other hand, quickly brought the oil supply down since Iraq virtually stopped producing and exporting oil. The importance and size of the country's share of the world oil production and oil reserves created fears of persistent undersupply among oil consumers (although in the beginning OPEC covered the Iraqi share of the world supply) and soon translated into a strong oil price rise.

Later in the decade, other factors distorted the oil market and oil price respectively. Katrina and Rita hurricanes that hit the US and some of its neighbouring countries in the summer of 2005 added to the already high pressure on the world oil supply and price. It was not just the effect of these natural disasters but also the corresponding activities on the oil market (i.e. speculation) that assisted for the oil price developments.

Iran nuclear tensions once again reinforced the fears of another conflict in the Middle East with potential serious consequences. The UN sanctions against the 5th biggest oil producer in the world at that time clearly signalled to the oil market participants an expected increase in oil prices, which proved to be not just a momentous phenomenon.

During the past couple of years, a few developing countries had remarkable rates of economic growth of their economies (double-digit for China, India and certain Middle East countries) which was also reflected in their demand for oil. Since the expectations for their growth and respective oil demand was often underestimated, quite often the producers could not increase supply to the same extent, which further distorted the market equilibrium and pushed the oil price up. The tendency shifted in 2008 when the looming of the world economic crisis caused a strong contraction of demand for the commodity.

Since a couple of years ago the oil market observed the highest volatility in its history. 2008 has also been an amazing year that illustrates the complexity of the market. In less than 6 months the price of oil dipped more than $100 per barrel – the largest decrease in such a short period (the price of the benchmark WTI in January 2009 was around 28% of its value in July 2008 – Chart 11). This signals that the high price was largely inflated and artificial and could not be sustained, at least not in the near future.

Chart 11

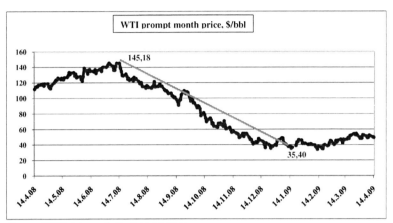

Source: ICE Database, Author's elaboration

The reality in 2009 was characterized by the world economic crisis and the solutions that were being sought and employed in order to resolve it. As most of the industries and markets, oil was also strongly affected and naturally the demand for the commodity started falling every day. Many were expecting the recovery of the economy and resurgence of the oil price. However, with the contradictory data about the state of the world economy that was published every day, it was too ambitious to make predictions about the eventual end of the crisis.

After pointing out the main events that influenced the oil market in the current decade, I continue with a closer look at the crude oil price term structure during this period. Chart 12 represents a time series of the term structure of oil futures prices for the period March 2003 – May 2008. The red line depicts the evolution of the spot crude oil price, while the dashed curves are the futures curves that show the price term structure at different points in time within the above-mentioned period.

Chart 12

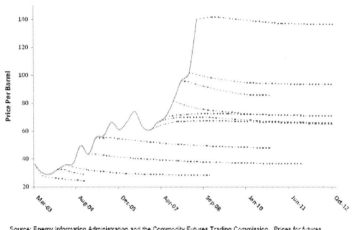

Source: Energy Information Administration and the Commodity Futures Trading Commission. Prices for futures contracts shown are for liquid markets (1000 or more contracts).

The above chart clearly shows that during this period of steadily rising oil price, its term structure was mostly in backwardation, with a few exceptions. The chart also supports the position that rising demand (faster than the supply), war conflicts and the threat of such, natural disasters all put pressure not only on the oil price but also on its term structure and provide a good "environment" for backwardation.

On the other hand, the world economic crisis, slowing of the growth and oil demand, a period of (seemingly) absence of major conflicts involving major oil producers and fierce natural disasters to distort oil supply, naturally brought a shift from the lasting backwardation to a steady contango (Chart 13 below). It characterizes the state of the economy quite well and reflects the severity of the situation. But there is something more to all that.

Chart 13

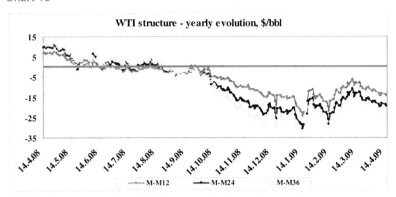

Source: ICE Database, Author's elaboration

So far I somehow ignored the main reason for the everyday volatility of the oil price and oil price term structure and the time has come to dig into it. As mentioned in the factors that influence the price term structure, futures' trading is among the most important causes for its evolution. And while I talked about this issue in a rather general manner, this has become an extremely important phenomenon only in the last decade. Why? It did not take long for the investment banks and other institutions to spot the tempting opportunities on the crude oil market exchanges and enter. The success of the first ones lured many more and trading oil futures has since become a regular part of the main activities of many financial institutions and investment funds. The volumes of these activities has increased so much that gradually they became responsible for setting the oil price. A paradox emerged as oil has become the only commodity in which its producers do not directly participate in the day-to-day price fluctuation. The reasons for which the oil futures were created have been heavily abused, which has transformed the oil market into a big mess. I agree with Mr. Hachfeld (the former head of the Market Analysis department of Litasco SA) that the arrival of "investors at-large" (non-commercial investors) has caused huge distortions in oil price curves and price structure. In support comes the following chart, representing the striking correlation between the volumes of non-commercial trade and the price of WTI crude oil:

Chart 14

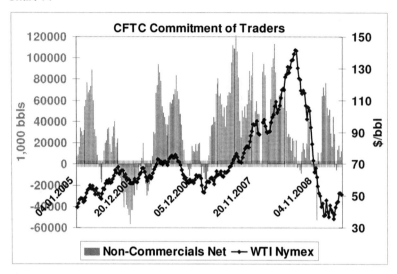

Source: Litasco SA database

The last table is striking for the way the price of WTI crude oil and the volume of non-commercial trading have moved during the last 4 years. In almost any period a relative decrease/increase in the volumes traded by non-commercial investors translate into a relative decrease/increase in the WTI crude oil price. And as this happens, the whole term structure is altered and updated according to the corresponding changes.

In conclusion, the last paragraph presented the main development of the crude oil term structure in the first decade of the 21st century. Although many other factors affected its evolution, the extent to which the activities of the non-commercial traders have influenced the industry and the market in the last five or ten years is remarkable and so far proves to continue to be the major day-to-day factor that would drive the crude oil term structure in the future.

6.2.5. Contango and backwardation revisited and the implications for the various crude oil market participants

The amazing picture in the above chart implies that oil price and oil price structure respectively have been largely driven by the volumes and positions of traders in futures. The culmination as seen above came in the summer of 2008 when the big bubble created by the huge amounts traded in "paper barrels" burst and the oil price started a free fall.

And it has not been just the oil price to be "locked in" by the non-commercial traders. Their activities have had an influence on all participants (from oil producers to refiners). By trading progressively larger volumes in crude oil for example, they have pushed up the front oil price by stimulating the creation of a backwardation structure. What have been its implications? Oil producers were content because they enjoyed higher spot prices, sustained by the backwardation. Refiners on the other hand started using their stocks since they became cheaper to use than buying from the market. As the normal stock cover of refiners is around 50-55 days, they hoped their lower demand would bring the price down. Traders adjusted their strategies and switched short positions with long in order to profit from the spread trade.

In backwardation, the market for physical (spot) oil is extremely rigid since anybody can enter into a futures contract for a delivery a few months ahead. Therefore, trade is commenced mainly on the futures exchanges.

As for the contango, while it has been wide enough to stimulate storage and sale in the future, a narrower contango leads traders to switch their strategy accordingly, by changing their positions. Having taken long positions before, they switch to short ones since the contango has become unattractive. Refiners are able to buy from the market, keep or even increase/rebuild their stocks and assure a comfortable margin. Producers may be negatively affected by a depressed oil price (if the contango is very wide) but nevertheless few of them participate on the futures exchange markets (e.g. the vast majority of OPEC members do not use futures). However, they may cut production if the price is too low in order to stabilize the oil price and the market.

With this I conclude the historical overview of the oil price term structure and move to the analytical part of this work. So far I have shown the main factors that influence the crude oil market and their implications, the main characteristics of the oil price term structure and the evolution of crude oil market throughout the years. This was important in order to present and explain the arguments that follow as clearly and objectively as possible. Having said this, I continue with the analytical study of crude oil price term structure.

7. Analytical study of crude oil price term structure

Although oil futures and oil market exchanges are still quite young, their relevance for the whole oil industry has stimulated the creation of various statistical databases and techniques to store and analyze information from the oil market. For example, ICE provides historical price data for more than 15 futures products with diverse maturities (up to 10 years ahead) for free on its website. And I gratefully borrowed the inputs for the analytical study that I would carry out next.

7.1. Methodology

Initially, I gathered pricing data for Brent and WTI – the prompt months and 3 additional futures prices with maturities of 1, 2 and 3 years from the prompt month. I have done the same for 3 oil products – ICE Gasoil, Nymex RBOB and ICE Heating oil as well as for another energy commodity – Henry Hub Natural Gas. The collected data is for the period starting from 02.01.2008 until 30.04.2009. Having this information, I was able to take the differential between the prompt month futures price and the other 3 futures prices with the respective maturities of M12, M24 and M36 from the prompt month. After that I had to ensure that all data is in the same unit and consequently I did some conversions for part of the data (for example, metric tons in barrels). All this enabled me to graphically build the evolution of the term structure for both crude oil and oil products for the period mentioned above. I decided to represent the data on a weekly, a monthly and an yearly basis in order to be able to consider short term but also longer term perspectives. Additionally, I took the differential of the results obtained from the previous operation: for example (prompt month – 1 year futures) – (prompt month – 2 year futures). The idea was to show the corresponding change in the evolution of the term structure day–by-day, for the same time basis.

7.2. Reasoning

I have concluded that in order to successfully track the evolution of the crude oil term structure throughout the years, the analysis had to start even before the creation of oil futures and oil futures markets. The importance of oil has grown incredibly throughout the past decades as well as the factors that affect its availability and price. I have reasoned that 2008 and 2009 represent everything that could be seen in the oil industry and market – a large variety of factors drove the market and oil price to extremities – and therefore I have concluded that it would be valid to restrict the analysis to this time period. I have decided to use the price of WTI crude oil as the basis for the work since it has been established as one of the most important benchmarks in the industry and since it is being quoted as the primary indicator of the world oil price in every respected media. However, the conclusions and results are equally valid also for Brent crude oil since there is a very strong correlation between the two leading benchmarks. Although the data for the 16 months that I compiled for the purposes of this work is enough to show the strategic implications of crude oil term structure, the historical timeline and the factors that influence the oil market and oil term structure are indispensable to the completeness and credibility of the analysis.

I have come to the conclusion that building the term structure of crude oil can bring (sometimes even unexpected) advantages for the professional reader and its correct analyzing could give a company or an individual trader a competitive edge on the market.

7.3. Results

The yearly results for WTI are represented in Chart 15:

Chart 15

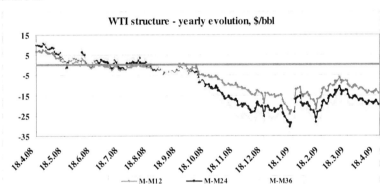

Source: ICE Database, Author's elaboration

The chart shows the term structure was in steady backwardation throughout the first half of 2008. The following 2-3 months it wandered between narrow contango and backwardation only to enter a widening contango with the looming of the world economic crisis. In the beginning of 2009, there was a situation of supercontango, which did not persist for long. The contango of futures with shorter maturities (M12) tended to widen at a slower pace with respect to the futures with longer maturities (M24 and M36). In the April 2009, the term structure somehow stabilized and was not as volatile as in the previous half an year. The differentials for the various futures also moved with a similar "tempo". Chart 16 below represents the monthly development of the WTI term structure.

Chart 16

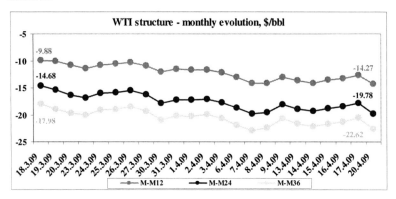

Source: ICE Database, Author's elaboration

7.4. Use of results for strategic decision-making

For most of the oil market participants today, spot oil price taken alone is no longer a meaningful indicator of the market. What matters much more today are concepts like basis and spread, or simply the crude oil price term structure.
After presenting the data and the results, I focus on what it actually implies for the strategic decisions of the different categories of market participants. In this connection, I first define the different categories of market participants, which I overlooked so far. I can extract them from the simplified value chain of the oil industry.

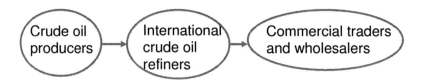

First, there are the oil producers. They, in addition could be further segmented to OPEC producers and non-OPEC producers, signalling the importance of the OPEC as a consolidated group of countries, which had the potential of adjusting their supply on a large scale and which have committed to keeping the oil market stable, that is to meet supply and demand.

OPEC oil producers however, do not participate in the oil market exchanges and they do not trade in futures (one reason is because they do not want to be seen as manipulating international oil prices) – their main market is the spot one, selling whatever they produce at the prevailing market price at that time. Their profit comes from the sales revenue at the spot market minus their costs of production. Therefore, they do not use the oil price term structure in their strategic decision-making process.

On the other side there are the non-OPEC oil producers. The majority of them, similarly to OPEC countries, do not participate in the oil futures markets – many of the non-OPEC countries use the oil they produce for their own needs (i.e. on average, US production was 8.49 million barrels per day in 2008 while in the same period its consumption was 19.42 million barrels per day). Yet most of the others, which are net exporters of crude oil, sell on the spot market. Moreover, since there is no coordination between those countries and each decides how much to produce based solely on its own needs and policy, they often go against the market logic and fundamentals, by not taking into account the signals from the market. The following chart illustrates this issue:

Chart 17

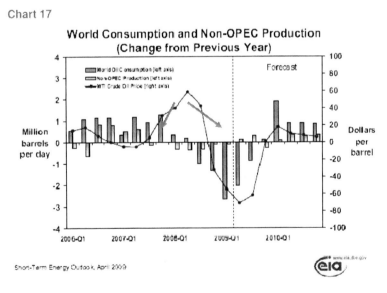

As could be seen from the pointed regions in Chart 17, often the non-OPEC producers have assisted in the increased volatility of the market and have acted against supply and demand fundamentals. In the end, they also rarely use price term structure in their strategic decision-making processes. However, they do influence it by their supply decisions and activities.

Proceeding along the value chain, we have the refiners. They could broadly be divided into three categories. The first one is represented by the fully-integrated refiners like Exxon Mobil, Shell, BP, ENI and a few others. They are characterized as part of completely vertically integrated oil companies with global activities all over the industry value chain. The second category is formed by "semi-integrated" refiners such as Repsol, Lukoil and others. They are similar to the previous group except for the extent of their activities along the value chain, expressed in less international presence in certain parts of the value chain (i.e. very geographically concentrated downstream (retailing) activities. The last category is represented by trading companies, which possess refineries and/or ships and/or storage capacities. Examples could be Glencore, Vitol, Taurus, and Litasco SA.

On a broad level, they can all be classified to a smaller or larger extent as commercial traders. That is because refining goes hand in hand with trading and refining could not be profitable if it is not complemented by continuous trading activities. The main characteristic of the commercial traders (and what strongly distinguish them from the non-commercial ones) is that they get hold and constantly deal with physical oil - that is all their activities, including those on the futures markets are subject to the fact that they actually own crude oil. Their main source of revenue comes from the sales of refined products minus the cost of the feedstock and refining. Their main use of the oil futures market is for hedging purposes in order to ensure a minimum price for their output. They rarely if ever engage in speculative activities. Refiners face serious strategic challenges since they are in between the oil producers and the traders and consumers of oil products.

What are these strategic challenges?

Revisiting the time gap issue that I discussed before, I go into details of the activities of most of the refiners. I mentioned before that they have and do decide on what inputs to purchase now based on futures prices and therefore price term structure 2-3 months ahead. However, once they have chosen what to purchase, it is not the end. By the time the crude oil they have ordered arrives, the market conditions could change a lot and what they have ordered may have become unneeded. Furthermore, their original strategy could have become obsolete and potentially leading to losses or lower margins than originally expected. Therefore, refiners need to be or to have trading capacity and experience in order to be flexible and react to changing market conditions. Therefore, buying the inputs they need is a continuous process and very often they do not just wait for the original order to arrive. Refiners instead engage in a continuous trade, which is in line with the market fundamentals and needs. They can sell the cargo, which have ordered before it actually arrives if market conditions change. And this involves substantial strategic decision-making capacity since refiners have to decide whether to switch to refining other crude grades, when to do it, or they can even decide not to refine if their margins go beyond a certain level. Instead they can engage in trading crude oil and oil products by using their price term structure once again. I will illustrate this with a recent example. In the end of April 2009, swine flu epidemic fears gripped the world. This was immediately reflected in the crude oil price. Not surprisingly, it fell more than 3% on the first day of trading after the announcements. This was largely attributed to expectations of further negative effects on the unstable world economy. More precisely, one of the most obvious reasons was the big decrease in demand for flying and travelling throughout the whole world, since cases of this flu were reported on virtually every populated continent. This had implications for the refiners as they had to revisit their strategies for producing jet fuel/kerosene and to adjust both their inputs and outputs accordingly, trading part of the crude oil they ordered for producing jet. And in reality, jet fuel fell by more than the overall decrease of the oil market with 10$ per tonne.

On what information this strategies and strategic decisions are based?

This is another very sensitive question since the volatility of the crude oil market is almost impossible to predict. Market participants do rely on market fundamentals – for example, observing and considering the factors I mentioned in section 5. However, many of these factors are extremely difficult to predict accurately and yet, for many of them it is simply impossible to do this. Consider the example of swine flu above. Nobody could have known about the appearance of this epidemic and respectively its effects. But it happened anyway and market participants had to adjust their strategies post factum. And again relying on the price term structure, they noticed the switch in the market preferences and started disposing of heavier crude grades (from which kerosene\jet fuel is refined) in favour of lighter ones.

Finally, I point out to the last group of market participants I consider in this work and that are the most active in the formation and use of the crude oil price term structure. I am referring to the non-commercial traders. Here again I should divide them in sub groups according to their activities and goals, if I want to be accurate.

Broadly we can distinguish between commercial and non-commercial traders. While I already qualified the commercial traders, I should revisit the main differences with the non-commercial ones and focus on their goals and strategies they employ to achieve them.

Commercial traders usually possess ample storage capacity and have the ability to increase or decrease it when needed. The main purpose of the trading activities of these companies is hedging. It is a crucial task for them to assure a minimum price for their crude oil or oil products. In this way commercial traders are able to control their margins and hedge the risk of the huge volatility of the oil market. They could also be classified as "tangible traders" since they actually get hold of the crude oil.

I left out on purpose one group of commercial traders since they are very different from the ones mentioned so far – this is the group of the final consumers. Airlines, power-producing and transportation companies are among the representatives of this group. It is easy to see why they form part of the commercial traders and why they are relevant – they deal with physical oil, use the futures market for mainly hedging purposes and hardly ever engage in speculation. That is why I include them in the broad group of commercial traders.

The strategies of the commercial traders are largely similar among the different categories. Their purpose is to hedge and offset their physical positions, which means to take the opposite position on the futures market to the one on the spot (physical) market.

On the other side of the spectrum, we have the non-commercial traders. Here we have a broad group of market participants ranging from investment banks and other financial institutions (i.e. Morgan Stanley, J.P. Morgan are among the most active) to hedge funds and investors-at-large.

In contrast to the commercial traders, the non-commercial traders (or "paper") usually lack any storage capacity. They rarely if ever take possession or deliver physical oil (hence "paper barrel" traders) and operate solely on the oil futures exchange markets. They are also broadly referred to as speculators since very often their activities deliberately target a distortion of the market, from which they profit by taking the appropriate market position. This leads us to their broad strategic means – speculating and betting on the market moving in one direction or the other which, if they were correct, brings them big profits. And of course, they do not have any crude oil to hedge.

This last point has the most important strategic implication, which is the key to understanding the functioning and strategic decisions of the various oil market participants. Having to deal with physical oil tremendously increases the complexity for the commercial traders and furthermore subjects them to further restrictions – both purely practical and regulatory (i.e. need to consider storage tariffs, freight rates, various countries' administrative requirements for delivering crude oil and oil products, etc). In the same time, the non-commercial traders have the flexibility and freedom to deal with as much "paper" barrels as they want. Consequently, the tools in their hands are much more diverse, which translates into a much wider array of strategic options available to them. Their overall expectation from the trade and speculation they initiate is bearing risk (sometimes very high) in return for possible profit (sometimes also very high). Therefore the stakes at hand are much higher. The problem is that on the overall the market is disproportionately affected by the successful execution of the non-commercial traders' strategies. If they gain, this has little effect on the oil market as a whole, while if they lose this usually distorts the oil price and market fundamentals and is negatively reflected in the results of all participants.

I distinguished and commented on the various groups of oil market participants and showed their interests and strategic activities on the market. The fast development of the oil market brought many new players, which somehow paradoxically succeeded to become so relevant, as to be able to exert serious influence on price and term structure. And there are no indications that this would change in the near future.

8. Relationship between crude oil, crude oil products and other energy commodities

There is a logical relationship between crude oil and crude oil products. As the former is the build stock for the latter, it is obvious to expect that as the price and availability of the input changes, similar effects are observed also for the output. But quite often in the oil industry, the reverse is also true – that is, a change in price and/or availability of oil products leads to corresponding changes in price and availability of crude oil. Price term structures are also highly correlated. Even more interesting is the analysis of the relationship between crude oil price and the price of other energy commodities. A notable example is the one of crude oil and natural gas. For many reasons we actually find a very high correlation between the two prices. In order to dig deeper into these issues, I would make one examples of the correlation between the price of crude oil and the price of one of its products and a second one dealing with the relationship between crude oil and natural gas.

8.1. The example of ICE Gasoil

Gasoil is a product of crude oil, obtained through the process of distillation. The oil is heated to a certain temperature until it becomes gas and later condenses. It is most commonly referred to as diesel fuel, used mainly in motor vehicles but also for heating. As it is among the most important oil products used by modern man, gasoil is traded on the ICE market exchange in great volumes and various factors affect its price and availability.

It is rather clear how crude oil "drives" gasoil. Crude oil supply disruptions for example push its price up, which often automatically translates into higher costs for refiners as this is the primary input for their production. In principle, absent any special events, any change in the price of crude oil should be equally affecting the price of gasoil. And the following chart of the prompt prices for both crude oil and gasoil for the period 22.04. 2008 – 22.04. 2009 supports this statement.

Chart 18

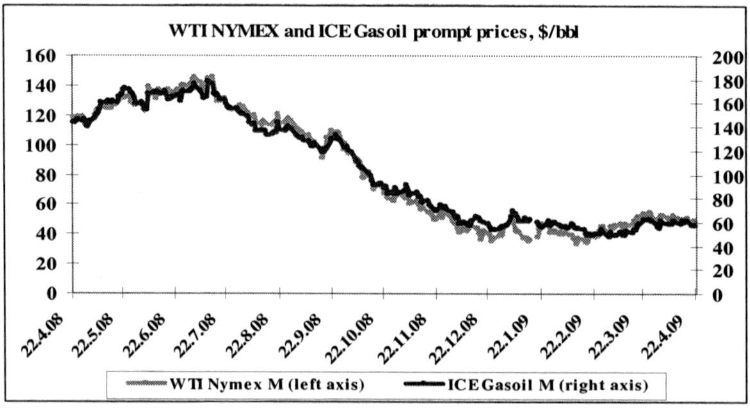

Source: ICE Database, Author's elaboration

Clearly both prices move almost perfectly together. And this is not something new or surprising. However, it is not only a change in the price of crude to trigger a change in the price of gasoil. Similarly to the example of the consequences after Katrina hurricane when speculative behaviour by gasoline retailers and the fear of the population about fuel availability put a heavy upward pressure on gasoline prices, changes in the oil products markets have often a strong potential to cause the crude oil price to move with them. For example, in theory a colder than expected summer temperatures cause the demand for gasoline and diesel fuels to be lower, which brings their prices down. This in turn puts pressure on the oil price and it goes down accordingly.

As one could have expected, there is a strong relationship between the prices of crude oil and crude oil products. In the next section however, I confront WTI crude oil price and the price of Henry Hub Natural Gas in order to check if they are related similarly as in the example above.

8.2. The example of Natural Gas (NG)

Natural gas can be referred to as an energy commodity that is often a substitute for crude oil. Similarly to crude oil, it should undergo a substantial processing before it is employed as a source of energy. Furthermore, natural gas is often associated with crude oil deposits as they are usually found together. Its uses are also similar and that is why it could be often considered a substitute product, subject to usage patterns and technology employed. It is a major source of power generation, has a wide application among households, and is a major feedstock for fertilizers and hydrogen.

Economic factors also link crude oil and natural gas and their prices respectively.

On the demand side, an increase in the price of crude oil leads consumers to change their preference for crude oil and switch to the use of natural gas. The rising demand for the latter pushes the prices up accordingly. However, there are limits to this substitution as crude oil and natural gas can be used interchangeably only in certain sectors (mostly in the electric generation and industrial sectors of the economy). According to EIA's Manufacturing Energy Consumption Survey (MECS), more than 18% of the usage of petroleum products can be switched to natural gas and vice versa.

Furthermore, if the price gap between NG and crude oil gets too large, consumers (especially industrial) start to switch between the two, which will eventually bring them again closer together.

On the supply side, a rising demand for crude oil, which is supported by an increase in oil extraction and production, may (often involuntary) increase the production of natural gas as a co-product of oil since many of the gas deposits are found and exploited together with those of crude oil – the so called associated natural gas. As a consequence, it is expected that under such circumstances the price of natural gas tends to decrease.

Under the same premises, we can observe a contrasting scenario. A rising crude oil demand may lead to increased costs of natural gas development and production and therefore pushing its price upwards. As crude oil and natural gas operators compete for rather similar economic resources like specialized labour and drilling rigs, an increase in the demand and price of crude oil would lead to higher levels of production activities as the operators would try to speed up the process and take advantage of the favourable market conditions. The bigger activity however, would bid up the costs of all of the relevant factors, which will therefore translate into higher costs of developing natural gas projects.

Another even more obvious linkage between crude oil and natural gas is the LNG or liquefied natural gas, which permits transportation of natural gas from large producing areas to large consuming areas and countries. And since most of the LNG contracts are indexed to crude oil and/or crude oil products, there is a direct linkage between prices of crude oil and natural gas.

Market behaviour implies that changes in the oil price lead to changes in the price of natural gas and not the other way around. An explanation for the asymmetric relationship is the relative size of the two markets. The price of crude oil is determined globally while natural gas markets are highly segmented and therefore events and conditions that influence the price of natural gas are unlikely to affect the global price of crude oil.

This was the theory that hinted that we should expect a correlation between the two products. What is the result in practice?

In a paper from Villar and Joutz (2006), they managed to statistically prove the aforementioned relationship. Using various tools (i.e. co-integration, vector auto regression) and applying them to the actual price data for the period January 1989 – December 2005, they found a strong correlation between the price of crude oil and the price of natural gas. Graphically:

Chart 19

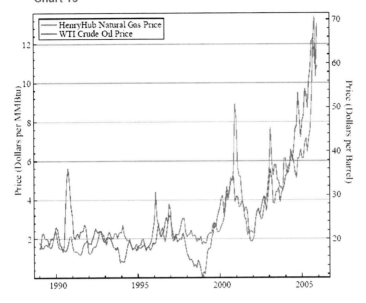

Source: Energy Information Administration, Short-Term Energy Outlook, various issues

From this chart is it obvious that very often a spike in the price of crude oil is reflected immediately in the price of natural gas. Statistically, this looks like the following regression chart:

Chart 20

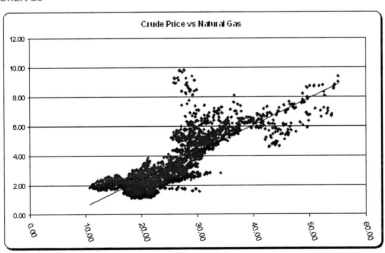

Source: Jason Braswell, Director of Risk Management Services Intelligent Energy®

Comparison of the 5-year futures prices for NG and WTI contracts on NYMEX also exhibit similar results.

Chart 21

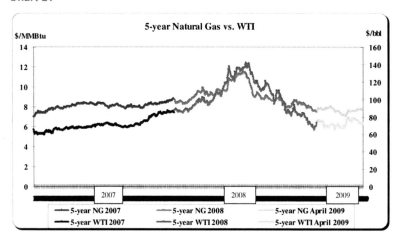

Source: ICE Database, Author's elaboration

In the last couple of years, an important factor emerged as an explanation to the continuous correlation between the prices of NG and crude oil, which was not so much relevant in the past. The booming participation of various commodity hedge funds and investment banks on the energy markets has helped in keeping the two in line as many of them have invested and traded the energy sector as a basket. Therefore, a distortion in the energy market has since influenced both prices similarly to a smaller or larger extent.

Despite periods of vague correlation between the price of crude oil and natural gas, a rather stable relationship between the two is identified both statistically and economically. Clearly the correlation in this case is weaker in comparison to the case of crude oil versus a crude oil product, but it is still a significant one.

9. The price term structure, utilized as a tool for forecasting purposes

Crude oil price term structure, much like the market for securities, cannot in general be deliberately "beaten". Quite often market participants (of whom mainly speculators) take a decision based on feelings, personal beliefs or even pure chance. And sometimes they are successful – after all they can be either right or wrong and the market can go only in two different directions. However, when we consider using the futures prices, futures market exchanges and the resulting price term structures for forecasting purposes – it is probably smarter to have a second thought. They, of course, give us a lot of inputs for strategic decision making and shape our ideas and view for the future. But to believe that by looking at and closely following them we are able to predict in which direction the market would go is not accurate.

Why this is the case?

The main reason is that market (economic) fundamentals alone are not enough in explaining the functioning and development of the crude oil market. For example, a falling demand and/or a rising US dollar do not always mean that the price of crude oil will fall for sure. And there are enough examples to prove this.

One reason is that there are so many factors that influence the oil market (I mentioned only the main ones in the beginning of this work) and it is simply impossible to predict when and how they would actually affect the industry. Whether there would be a stronger than expected hurricane that would seriously distort the global production capacity; or whether a new type of flu would create a world panic and precautions, which is immediately reflected in the oil market. These are examples, which hit the oil market in the last decade and which supported the statement of its unpredictability.

One of the conditions why any major factor has such a serious influence on the whole industry is that crude oil is among the very few commodities which are truly global and are traded predominantly on global market exchanges. Therefore, no matter whether something affects only the US industry or the Russian production – if it is large enough, it would cause distortions to the global oil market.

The concentrated deposits of oil reserves around the world also contribute to the complexity of the market. For example, even minor information about possible war in Iran is immediately reflected in the oil price and term structure by putting an upward pressure, no matter whether the market fundamentals would suggest otherwise.

An even more striking example of the unpredictability of the oil market is represented by the spike in oil prices in the summer of 2008. In the 3 months of record high prices, the demand was actually falling and quite lower than supply (Table 3).

Table 3

	2008	
	1st	2nd
Supply (million barrels per day) (a)		
OECD	21.29	21.09
U.S. (50 States)	8.62	8.75
Canada	3.38	3.23
Mexico	3.29	3.19
North Sea (b)	4.47	4.33
Other OECD	1.53	1.58
Non-OECD	64.05	64.52
OPEC	35.66	35.83
Crude Oil Portion	31.25	31.40
Other Liquids	4.41	4.42
Former Soviet Union	12.59	12.60
China	3.94	4.00
Other Non-OECD	11.86	12.10
Total World Production	85.33	85.61
Non-OPEC Production	49.68	49.78
Consumption (million barrels per day) (c)		
OECD	48.68	47.09
U.S. (50 States)	19.88	19.68
U.S. Territories	0.27	0.28
Canada	2.37	2.25
Europe	15.20	14.89
Japan	5.41	4.59
Other OECD	5.55	5.39
Non-OECD	37.71	38.14
Former Soviet Union	4.35	4.30
Europe	0.83	0.79
China	7.74	7.99
Other Asia	9.22	9.26
Other Non-OECD	15.58	15.80
Total World Consumption	86.39	85.24

Source: Short-Term Energy Outlook, EIA, March 2009

Even a simple analysis of the price forecasts of the 2 major world agencies, dealing specifically with energy, show huge deviations, even above 100% in certain years.

For example, for the past dozen years the average error in the annual forecast of EIA, which is based on various analyses of market expectations, oil futures and price term structure, was 53%. In its Annual Energy Outlook of 2002, the same agency forecasted that the average oil price for 2008 would be $23.11/bbl.

Furthermore, as I discussed earlier that in the last years increasingly the world crude oil price has been produced by the huge volumes of trading activities on the oil futures market exchanges, this hints that the importance of physical oil for setting the world price has significantly diminished. Therefore we may have a stable supply and market for physical oil, but the trading in "paper" barrels is what should be observed as the main driver of the price. And since much of this trade is ultimately based on personal feelings and expectations without any strong basis for justification, it is practically impossible to have a valid forecast, based on relevant data.

10. Final remarks and conclusions

In this last section I summarize the main goals of the paper and the main points I tried to make. I started with a general overview of crude oil market and then went on to explain what is behind crude oil price term structure. Later, I pointed out the main factors that influence the price and term structure of crude oil since it was indispensable to support many of my later statements and ideas. Then I started hinting the main task of this work by discussing the importance of the two states of the term structure – contango and backwardation. The historical evolution of crude oil price, market and term structure represented how quickly they have developed throughout the years, what were the main drivers and events that led to market changes and most importantly why we are facing this crude oil market and industry today. The main goal has been to bring more light to the term structure and its implications for the various groups of crude oil market participants. I wanted to discuss all the issues that they face along the value chain and to point out whether and how they use the term structure and whether and how they influence it.

The relationship between crude oil, crude oil products and other energy commodities was presented to illustrate the interdependence between the various components of the energy industry and to give insights as to how to think about them.

Finally, we observed that crude oil price is among the most difficult things to predict and its term structure is definitely not among the tools that could help for this purpose.

In conclusion, I present the following table:

Table 4

	Before	Today
Price	Stable	Volatile
Participants	Limited	Numerous
Market Forces	Fundamentals	Fundamentals and Sentiments
Trade	Regional	Global
Type of Business	Physical Only	Paper (Derivatives = 10/15x Physical Volumes)

It summarizes the main differences between crude oil market in the past (up to 20 years ago) and the one of today. All of it has been discussed in the paper in detail as to show how the transition actually happened and what the consequences were for the various market participants. There are no obvious indications that anything will change soon on the crude oil market but as I tackled before a few of its obvious malfunctions and problems, it would be no surprise to observe the next stage of evolution rather sooner than later.

References:

Litasco Database and Documents

Energy Information Administration Database

International Energy Agency Database

Reuters Database

Platts Database

Mileva Elitza and Siegfried, Nikolaus "OIL MARKET STRUCTURE, NETWORK EFFECTS AND THE CHOICE OF CURRENCY FOR OIL INVOICING", European Central Bank, Occasional Paper Series, NO. 77, December 2007

"Understanding Today's Crude Oil and Product Markets", A Policy Analysis Study by Lexecon, An FTI Company, 2006

Gabillon, Jacques "The Term Structures of Oil Futures Prices", Oxford Institute for Energy Studies, WPM17, 1991

CERA, "Low Oil Prices Putting Supply Growth at Risk", 2009

Parsons, John E. "Oil Prices", Center for Energy and Environmental Policy Research, December 12, 2008

Dale, Charles and Zyren, John, "Noncommercial Trading in the Energy Futures Market", Petroleum Marketing Monthly, May 1996

Hamilton, James D. "Understanding Crude Oil Prices", Department of Economics University of California, San Diego, May 22, 2008, Revised: December 6, 2008

152nd (Ordinary) Meeting of the OPEC Conference, Press Release No 4/2009, Vienna, Austria, 15 March 2009

New York Mercantile Exchange, "A Guide to Energy Hedging", 2008

Tamny, John "Oil, the Dollar and Comparative Advantage", CBOE Futures Exchange publication, July 8, 2008

http://www.wtrg.com/prices.htm

Gartman, Dennis "Understanding the Important Difference between Contango and Backwarded Forward Futures Prices", February 28, 2008

Langley, Charles "Paper money chasing paper barrels: Why oil prices are artificially high", June 10th, 2008

Lin, C.-Y.C. (2009). Insights from a simple Hotelling model of the world oil market. Natural Resources Research, 18 (1), 19-28.

Zhang, Yue-Jun & Fan, Ying & Tsai, Hsien-Tang & Wei, Yi-Ming, 2008. "Spillover effect of US dollar exchange rate on oil prices," Journal of Policy Modeling, Elsevier, vol. 30(6), pages 973-991.

Hanke, Steve H. "Weak Dollar and US Petroleum Reserves Behind Strong Oil Price." June 25, 2008. http://www.cato.org/pub_display.php?pub_id=9490

Villar, Jose A. and Joutz, Frederick L. (2006). " The Relationship Between Crude Oil and Natural Gas Prices", Energy Information Administration, Office of Oil and Gas, October 2006

JBC Energy, Market Watch: Special Oil Market Analysis and Outlook, March 2009

Braswell, Jason (2005), "Wrestling with Chaotic Natural Gas Prices", Daily News, Tuesday, April 12, 2005

Harper, David (2007), "Contango Vs. Normal Backwardation", Investopedia, 2007

Darrat, Ali F, "On the role of futures trading in spot market fluctuations: perpetrator of volatility or victim...", Journal of Financial Research, September 2002

Interagency Task Force on Commodity Markets, Interim Report of ITF, July 2008

Zeal_LLC, "Soaring Gasoline, Diesel, and Crude Oil Prices", May 09, 2008

Fattouh, Bassam, (2006), "Contango Lessons", *Oxford Energy Comment*, Oxford Institute for Energy Studies, September 2006. Published in *MEES* 49:48 dated 27 November 2006.

OPEC, *OPEC Monthly Oil Market Report*, March, 2009.

Gorton, Gary B et al (2007), "The Fundamentals of Commodity Futures Returns", Yale ICF Working Paper No 07-98, July 2007

Saxton, Jim, US Congress Joint Economic Committee, "OPEC's 902 Billion Barrel Oil Reserve", Research Report #109-28, January 2006

David Pumphrey, Center for Strategic and International Studies, "Outlook for Oil Market", 2009 NASEO Energy Outlook Conference, February 2009

Barclay's Capital, "Oil Sketches", November 2008

G20 London Summit Final Communiqué, "The Global Plan for Recovery and Reform", 2 April, 2009

SCOTT MACLEOD/DOHA, "Oil Prices: Don't Blame OPEC", October 2007

Rapier, Robert "The Energy Information Providers: EIA, IEA and CERA", The Oil Drum, ASPO 2008 Sacramento, September 21, 2008

Weinschenk, Matt "Contango: The Most Profitable "Buy-and-Hold" for 2009", Senior Analyst, White Cap Report, 2009

Tchilinguirian, Harry and Mortimer, Paul Lee, "What are Oil Futures and How they are Traded?", BNP Paribas, 30 May 2008

Bamberger, Robert L. and Kumins, Lawrence "Oil and Gas: Supply Issues After Katrina", CRS Report for Congress, September 6, 2005

Alhajji, A.F. and Huettner, David "OPEC and World Crude Oil Markets from 1973 to 1994: Cartel, Oligopoly, or Competitive?", The Energy Journal, June 2000

Asche, Frank; Gjolberg, Ole and Volker, Teresa "Price Relationship in the Petroleum Market: An Analysis of Crude Oil and Refined Product Prices", Foundation for Research in Economics and Business Administration, Bergen, August 2001

Kumar, Manmohan S. "The forecasting accuracy of crude oil futures prices", International Monetary Fund Staff Papers, JUNE 1992

Chantziaraa, Thalia and Skiadopoulosb George "Can the Dynamics of the Term Structure of Petroleum Futures be forecasted? Evidence from Major Markets"
First Draft: 1/10/2005 – This Draft: 6/12/2005

Kogan, Leonid; Livdany, Dmitry and Yaronz, Amir "Oil Futures Prices in a Production Economy with Investment Constraints", FORTHCOMING JOURNAL OF FINANCE, April 2008

Scientific Publishing House
offers
free of charge publication
of current academic research papers, Bachelor´s Theses, Master's Theses, Dissertations or Scientific Monographs

If you have written a thesis which satisfies high content as well as formal demands, and you are interested in a remunerated publication of your work, please send an e-mail with some initial information about yourself and your work to *info@vdm-publishing-house.com*.

Our editorial office will get in touch with you shortly.

VDM Publishing House Ltd.
Meldrum Court 17.
Beau Bassin
Mauritius
www.vdm-publishing-house.com

CPSIA information can be obtained at www.ICGtesting.com
Printed in the USA
LVOW130219240613

339906LV00001B/83/P

9 783844 324877